Remodelers' COST OF DOING BUSINESS STUDY

2023 Edition

NAHB®
National Association of Home Builders

Remodelers' Cost of Doing Business Study, 2023 Edition

BuilderBooks, a Service of the National Association of Home Builders

Patricia Potts	Senior Director
Design Central	Cover Design
Robert Brown	Composition
Gerald M. Howard	NAHB Chief Executive Officer
Robert Dietz	Senior Vice President and Chief Economist, Economics and Housing Policy
John McGeary	Senior Vice President, Business Development and Brand Strategy
Paul Emrath	Vice President, Economics and Housing Policy
Rose Quint	Assistant Vice President, Survey Research, Economics and Housing Policy

Published in the United States of America

26 25 24 23 1 2 3 4 5

ISBN-13: 978-0-86718-802-8
eISBN-13: 978-0-86718-803-5

For further information, please contact:
National Association of Home Builders
1201 15th Street, NW
Washington, DC 20005-2800
BuilderBooks.com

Contents

Figures

Chapter 1. All Residential Remodelers

Chapter 2. General Remodelers

Chapter 3. Design-Build Remodelers

Chapter 4. Comparison Across Regions, Business Models, & Historical Trends

Chapter 5. Operations

Tables

Chapter 1. All Survey Respondents

Chapter 2. General Remodelers

Chapter 3. Design-Build Remodelers

Chapter 4. Comparison Across Regions, Business Models, & Historical Trends

1

All Residential Remodelers

Prior to this study, NAHB had conducted similar analysis on industrywide profit margins for residential remodelers in fiscal years 2003, 2011, 2015, and 2018. The latest *Remodelers' Cost of Doing Business Study* asked for financial statements for fiscal year 2021 (income statement and balance sheet), plus additional information regarding the specific type of remodeling company, the number of employees on payroll, the number of jobs completed in different price ranges, and years in business. The survey questionnaire was sent electronically to about 5,500 NAHB remodeler members in April 2022. A total of 228 responses were received, but not all respondents answered every question.

Results are tabulated for all residential remodelers as a single group, and when possible, broken down separately by region of the country (the Northeast and South had insufficient responses to produce reliable estimates of financial performance, for example). In addition, results are tabulated by type of residential remodeler (general vs. design-build), total 2021 revenue, number of jobs completed in 2021, and number of years in business. Table 1.1 shows the distribution of responses across these groups.[1] The study also analyzes results for the most and least successful group of remodelers, namely those in the top 25th percentile and bottom 25th percentile, respectively, in terms of net profit margins. The financial analysis presented throughout this study is based solely on responses from members who reported their main operation to be residential remodeling/rehabilitation.

Table 1.1. Respondents' Profile

Region	**% of total**
Northeast	10
Midwest	24
South	46
West	19
Type of Residential Remodeler	
General remodeler	42
Design-Build remodeler	51
Specialty contractor	4
Repair & maintenance contractor	3
2021 Revenue	
Less than $2 million	64
$2 million or more	36
Number of Jobs	
30 jobs or fewer	53
More than 30 jobs	47
Number of Years in Business	
20 years or fewer	42
More than 20 years	58

[1] Due to rounding, the numbers in the tables presented throughout the study may not add up to 100 percent.

A plurality of the study's respondents operates in the South region of the country (46%), while 24% come from the Midwest, 19% from the West, and 10% from the Northeast. The regional distribution of those who provided a valid income statement for analysis, however, is significantly different: 42% of them come from the Midwest, 28% from the West, 22% from the South, and 8% from the Northeast. In the end, the number of responses from the South and Northeast were insufficient to produce reliable estimates of average profit margins separately for those regions.

Ninety-three percent of residential remodelers responding to the survey are either design-build remodelers (51%) or general remodelers (42%). Only 4% are specialty contractors and 3% are repair/maintenance contractors. In terms of total revenue in fiscal year 2021, 64% reported earning less than $2 million while the remaining 36% earned $2 million or more.

Looking at the number of jobs completed in 2021, 53% of responding remodelers finished 30 or fewer jobs while 47% finished more than 30 jobs. A breakdown by tenure in the industry shows that 42% have been in business for up to 20 years and the other 58% for more than 20 years (Appendix A).

Balance Sheet and Income Statement

The *Remodelers' Cost of Doing Business Study* survey asked remodelers to provide details about their 2021 fiscal year operations. Specifically, participants were asked for the amount of all the assets on their balance sheet (including cash, accounts/note receivables, and other assets), all liabilities (current and long-term), as well as the amount held as owner's equity.

Additionally, and to calculate industrywide averages for gross and net profit margins, respondents were asked for data from their 2021 income statements, including various lines of revenue, costs of sales, and operating expenses. In terms of revenue, the survey asked for the amount earned from residential remodeling/repairs, commercial remodeling/repairs, and new single-family home building. It asked about various costs of sales such as labor, materials, and trade contractors for residential remodeling/repairs, as well as direct construction costs for any commercial remodeling or new single-family home building and land. Data on various operating expenses were also requested, including indirect construction costs, financing expenses, sales and marketing expenses, general and administrative expenses, and owner's compensation.

To provide historical context for the current findings, this study compares the 2021 data with results from the previous two editions of the *Remodelers' Cost of Doing Business Study* (covering fiscal years 2015 and 2018). As table 1.2 shows, total revenue among residential remodelers responding to the survey in 2021 averaged $1.9 million, 15% lower than the amount reported in 2018 ($2.3 million) but 8% higher than in 2015 ($1.8 million).

Table 1.2. Gross Profit

	2015		2018		2021	
	Average ($1,000s)	Share of revenue (%)	Average ($1,000s)	Share of revenue (%)	Average ($1,000s)	Share of revenue (%)
Revenue	1,782	100.0	2,267	100.0	1,919	100.0
Cost of sales						
Labor for remodeling/repairs*	273	15.3	337	14.8	209	10.9
Materials for remodeling/repairs*	328	18.4	417	18.4	300	15.6
Trade contractors for remodeling/repairs*	462	25.9	650	28.7	682	35.6
Direct costs: commercial remodeling/repairs	38	2.1	13	0.6	12	0.6
Direct costs: single-family and land	96	5.4	68	3.0	178	9.3
Other costs	70	3.9	103	4.5	59	3.1
Total cost of sales	1,266	71.1	1,586	69.9	1,440	75.1
Gross profit	**515**	**28.9**	**682**	**30.1**	**479**	**24.9**

* Residential

All costs of sales combined averaged $1.4 million, accounting for a record 75.1% of revenue, up from 69.9% in 2018, and the highest share of revenue remodelers have reported spending on cost of sales since the inception of this study (see Chapter 4). Two cost lines stand out as having increased significantly from 2018 to 2021: the cost of trade contractors jumped from 28.7% of revenue to 35.6%; while direct construction costs and land for single-family home building tripled from 3.0% to 9.3%[2] during this period.

Subtracting all costs of sales from revenue shows remodelers' average gross profit in 2021 was $479,000, a 24.9% gross profit margin (479,000 ÷ 1,919,000). That average margin was about five points lower than in 2018 (30.1%) and, in fact, it is the lowest average gross profit margin remodelers have reported since the inception of this series in 1990. The previous lowest point was 26.8% in 2011.

While the share of revenue absorbed by the cost of sales skyrocketed from 2018 to 2021, operating expenses did the exact opposite. Rigorous belt-tightening on the part of remodelers led to an impressive drop in the share of revenue spent on these expenses, which went from absorbing 24.8% of revenue to 20.3%. As a result, remodelers in 2021 spent the smallest share of revenue on operating expenses in the history of the series (see Chapter 4).

The dramatic fall in expenses between 2018 and 2021 can be attributed to declines in two items: general and administrative expenses, whose share of revenue dropped from 12.6% to 7.2%; and sales and marketing expenses, which were cut in half from 3.2% to 1.6% **[Perspective from a Remodeler:** Most remodelers were swamped with calls without spending any extra to generate additional revenues.**]** (table 1.3). Other expenses either remained unchanged (owner's compensation stayed at 5.3% of revenue) or showed slight growth (indirect construction costs rose from 3.6% to 5.3%).

[2] Residential remodelers in 2021 were significantly more likely to be involved in single-family home building than in 2018. In fact, while only 5.6% of their total revenue in 2018 was derived from this secondary activity, the share rose to 11.0% in 2021.

Table 1.3. Net Profit

	2015		2018		2021	
	Average ($1,000s)	Share of revenue (%)	Average ($1,000s)	Share of revenue (%)	Average ($1,000s)	Share of revenue (%)
Gross profit	515	28.9	682	30.1	479	24.9
Operating expenses						
Indirect construction costs	92	5.2	81	3.6	102	5.3
Financing expenses	4	0.2	4	0.2	16	0.8
Sales & mktg. expenses	39	2.2	73	3.2	30	1.6
General & administrative expenses	188	10.6	285	12.6	139	7.2
Owner's compensation	97	5.5	120	5.3	102	5.3
Total operating expenses	420	23.6	563	24.8	389	20.3
Net profit (loss)	**95**	**5.3**	**118**	**5.2**	**90**	**4.7**

Figure 1.1. **Cost of Sales, Expenses, and Net Profit (% of Revenue)**

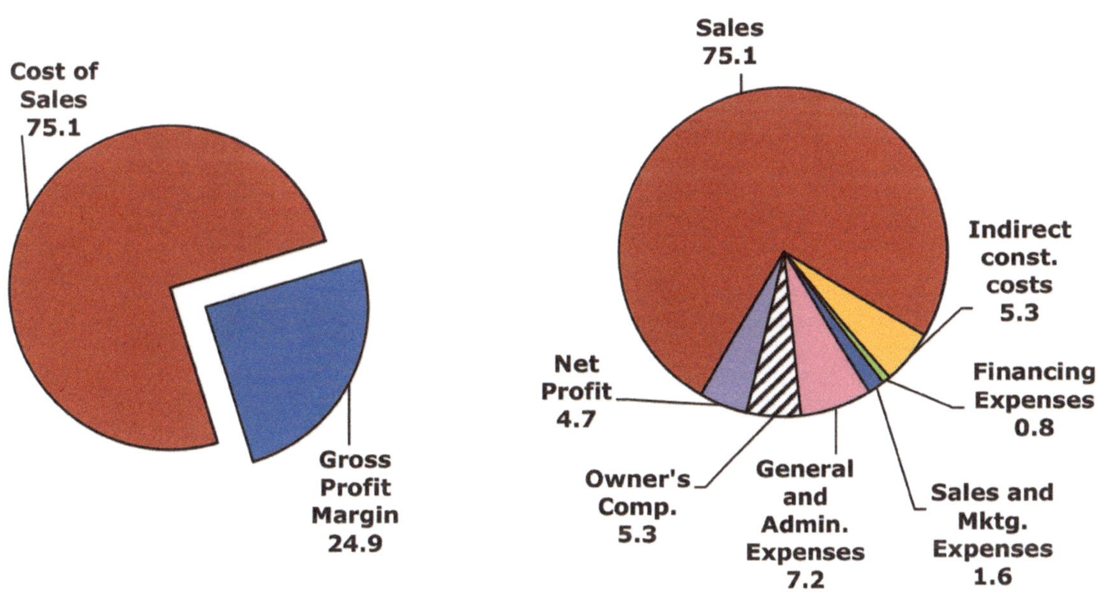

In the end, a deep contraction in total operating expenses was not enough to make up for the hit caused by higher costs of sales, and as a result, remodelers' average net profit margin was lower in 2021, at 4.7%, than in 2018 (5.2%). The last time remodelers' net margin was lower than in 2021 was in 2011, when it averaged 3.0%. Figure 1.1 summarizes remodelers' income statements for 2021.

In addition to the income statement, the *Remodelers' Cost of Doing Business Study* has tracked the balance sheet of remodelers over time. Results show that despite the slight retreat in net profit margin in 2021, remodelers managed to strengthen their balance sheets. As table 1.4 shows, survey respondents had total assets averaging $497,000 in 2021, 18% higher than in 2018 ($421,00) and 20% higher than in 2015 ($414,000).[3]

The plurality of remodelers' assets in 2021 was held as cash ($210,000 or 42.3% of total assets), another 22.7% ($113,000) was tied to accounts and notes receivables, and 13.9% ($69,000) was in the form of other current assets (such as refundable deposits or prepaid expenses). In all, 78.9% of remodelers' assets were current assets and the other 21.1% were fixed assets. It is worth noting that the share of assets held in cash has been growing in recent years: from 24.1% in 2015, to 39.5% in 2018, and 42.3% in 2021. **[Perspective from a Remodeler:** The current improvement may be a factor of the PPP (Paycheck Protection Program) loans that were later forgiven or EIDL (Economic Injury Disaster) loans taken out.]

Meanwhile, the other side of the ledger shows that remodelers are relying less and less on debt to run their businesses. Respondents owed an average of $242,000 in 2021, 10% less than in 2018 ($220,000) and 13% less than in 2015 ($279,000). The decline means that, for the first time in the history of the series, total liabilities financed less than half (48.6%) of all remodelers' assets in 2021. In 2015 and 2018, that share had been 67.5% and 52.3%, respectively. Current liabilities have been particularly curtailed in recent years, financing 47.5% of assets in 2015, 39.9% in 2018, and only 33.2% in 2021.

---

[3] The *Remodelers' Cost of Doing Business Study* is not designed as a longitudinal survey tracking the same companies over time. Instead, each year's study is based on responses from independent samples of current members of the NAHB.

Total revenue among remodelers in the top 25% group averaged \$1.5 million in 2021. They spent 62.6% of that on cost of sales, leaving \$558,000 in gross profits, a 37.4% gross profit margin. After subtracting operating expenses equivalent to 22.8% of revenue, the average net profit for this most successful set of remodelers was \$219,000, a 14.7% net profit margin (fig. 1.2).

Figure 1.2. **Profit Margins and Owner's Compensation (% of Revenue)**

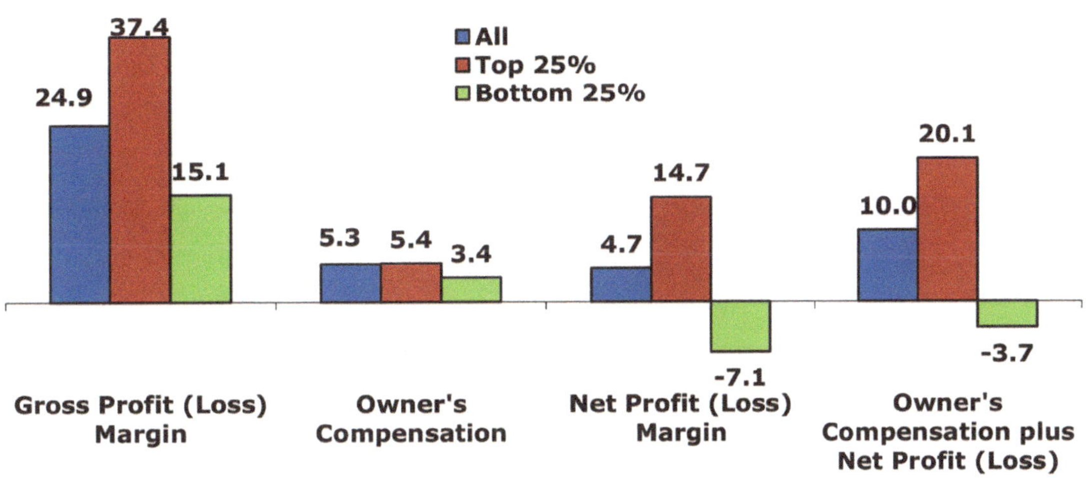

Remodelers in the bottom 25% group averaged \$1.6 million in revenue in 2021, of which an alarming 84.9% was spent on cost of sales. That left \$249,000 in gross profits, for a 15.1% gross margin (less than half the gross margin among the top 25% group). Operating expenses among this least successful group took up 22.2% of revenue, driving their net performance into negative territory: a net loss of \$117,000, or a net loss margin of -7.1%.

Because the line between the owner's compensation and the business' profit can be blurry in many companies (especially small ones), it is also useful to analyze a broader measure of performance that combines the share of revenue spent on owner's compensation and the net profit margin. As figure 1.2 shows, this overall measure of profitability was 10.0% for all remodelers combined, 20.1% for the top 25%, and -3.7% for the bottom 25%. In 2018, that measure was lower for the top 25% group (16.5%), but higher for the bottom 25% (5.6%). These results suggest that the severe disruptions caused by the COVID-19 pandemic to the economy and the housing industry may have contributed to improve the performance of the most successful remodelers, but also to further weaken the financial performance of the least successful ones.

Appendix B shows the average income statement for the top and bottom 25% groups side by side, allowing for a quick comparison of each cost and expense line between the two. A couple of items are particularly interesting. First, the most profitable group of remodelers in 2021 focused almost exclusively on residential remodeling, deriving 97% of their revenue from this activity and none of it from new single-family home building. In contrast, the least successful group endeavored in the construction of single-family homes, eventually accounting for 8.9% of their revenue in 2021. The second major difference in the two income statements is how much more the bottom 25% group spent on materials for residential remodeling work (19.2% of revenue), compared to the top 25% group (only 11.6%) (Appendix B, page B-2).

The 2021 balance sheet for remodelers in the top 25% group shows they had an average of $443,000 in total assets, $178,000 in total liabilities, and $264,000 in owner's equity. Remodelers in the bottom 25%, on the other hand, had slightly (6%) fewer assets ($418,000), 41% less owner's equity ($157,000), but 47% more in liabilities ($261,000). Table 1.6 summarizes important statistics from the income statement and balance sheet for the top and bottom 25% of respondents.

Table 1.6. Top and Bottom 25%

	2015		2018		2021	
	Top 25%	**Bottom 25%**	**Top 25%**	**Bottom 25%**	**Top 25%**	**Bottom 25%**
Gross profit (loss) margin (%)	35.8	23.1	31.4	29.8	37.4	15.1
Net profit (loss) margin (%)	16.8	(9.9)	13.3	(1.0)	14.7	(7.1)
Total assets (thousands $)	547	293	306	335	443	418
Owner's equity (thousands $)	292	24	172	161	264	157
Current ratio	3.35	1.10	3.73	1.95	2.76	1.96
Debt-to-equity ratio	0.87	11.34	0.78	1.07	0.67	1.66
Return on assets (%)	56.2	(46.8)	69.4	(5.9)	49.4	(28.0)
Return on equity (%)	105.3	(578.1)	123.4	(12.3)	83.0	(74.5)

The average gross profit margin among remodelers in the top 25% has remained above 30% since first calculated in 2003. It had declined from 2015 to 2018 (35.8% to 31.4%) but gained ground again in 2021 (37.4%). The net margin has fluctuated between 9.5% and 16.8% since 2003. Because their assets and equity rose at a faster clip than net profits between 2018 and 2021, this group's ROA (69.4% to 49.4%) and ROE (123.4% to 83.0%) both declined during this period. This means they still produced profits equivalent to 83% of their invested capital!

Findings for the bottom 25% group show a sharp deterioration in 2021. Their gross margin fell to an all-time low of 15.1% (led by large cost of sales related to single-family home building); while their net margin sank to -7.1%, worse than in 2018 (-1.0%), yet milder than in 2015 (-9.9%). Although the current and debt-to-equity ratios did not change significantly from 2018 to 2021, the loss in net margin led to declines in both ROA (-5.9% to -28.0) and ROE (-12.3% to -74.5%).

2

General Remodelers

Residential remodeling/rehabilitation is a broad industry with four major subcategories in which businesses can specialize in: general remodeler, design-build remodeler, specialty contractor, and repair & maintenance contractor. This chapter analyzes the financial performance of the 42% of respondents who identified as general remodelers.

Balance Sheet and Income Statement

On average, general remodelers reported \$1.8 million in total revenue for 2021, about 8% more than in 2018 (\$1.6 million) and about the same as in 2015 (\$1.8 million). They spent 83.0% of that revenue on cost of sales, a dramatic increase from the 72.9% spent on those costs in 2018. This left a gross profit margin of only 17.0%, compared to 27.1% in 2018 and 28.8% in 2015. The deterioration in general remodelers' gross margin was led by a sharp increase in the direct costs associated with building single-family homes, which went from taking up 1.4% of revenue in 2018 to 25.4% in 2021 (table 2.1).

Probably because of the significant reduction in their gross margins in 2021, general remodelers put operation expenses on a harsh diet and shrank them to account for only 15.3% of revenue — down from 22.2% in 2018. (table 2.2). Most strikingly, they managed to spend only 5.9% of total revenue on general and administrative expenses, about half of the 11.9% spent in 2018. In the end, despite the rigorous belt-tightening of expenses, the loss in gross margin caused general remodelers to report an average net margin of only 1.7% — less than half of what it was in 2018 (4.9%) and 2015 (6.3%). Figure 2.1 summarizes the income statement for general remodelers for 2021.

Table 2.1. Gross Profit

	2015		2018		2021	
	Average ($1,000s)	Share of revenue (%)	Average ($1,000s)	Share of revenue (%)	Average ($1,000s)	Share of revenue (%)
Revenue	1,812	100.0	1,642	100.0	1,778	100.0
Cost of sales						
Labor for remodeling/repairs	301	16.6	219	13.4	123	6.9
Materials for remodeling/repairs	292	16.1	289	17.6	221	12.4
Trade contractors for remodeling/repairs	425	23.5	538	32.7	641	36.1
Direct costs: commercial remodeling/repairs	19	1.1	22	1.3	26	1.5
Direct costs: single-family and land	118	6.5	24	1.4	452	25.4
Other costs	135	7.5	105	6.4	12	0.7
Total cost of sales	1,291	71.2	1,197	72.9	1,476	83.0
Gross profit	**521**	**28.8**	**445**	**27.1**	**302**	**17.0**

Table 2.2. Net Profit

	2015		2018		2021	
	Average ($1,000s)	Share of revenue (%)	Average ($1,000s)	Share of revenue (%)	Average ($1,000s)	Share of revenue (%)
Gross profit	521	28.8	445	27.1	302	17.0
Operating expenses						
Indirect construction costs	116	6.4	69	4.2	79	4.4
Financing expenses	2	0.1	3	0.2	1	0.1
Sales & mktg. expenses	23	1.3	22	1.4	24	1.4
General & administrative expenses	167	9.2	195	11.9	106	5.9
Owner's compensation	99	5.5	75	4.6	62	3.5
Total operating expenses	408	22.5	365	22.2	272	15.3
Net profit (loss)	**114**	**6.3**	**81**	**4.9**	**30**	**1.7**

Figure 2.1. **Cost of Sales, Expenses, and Net Profit (% of Revenue)**

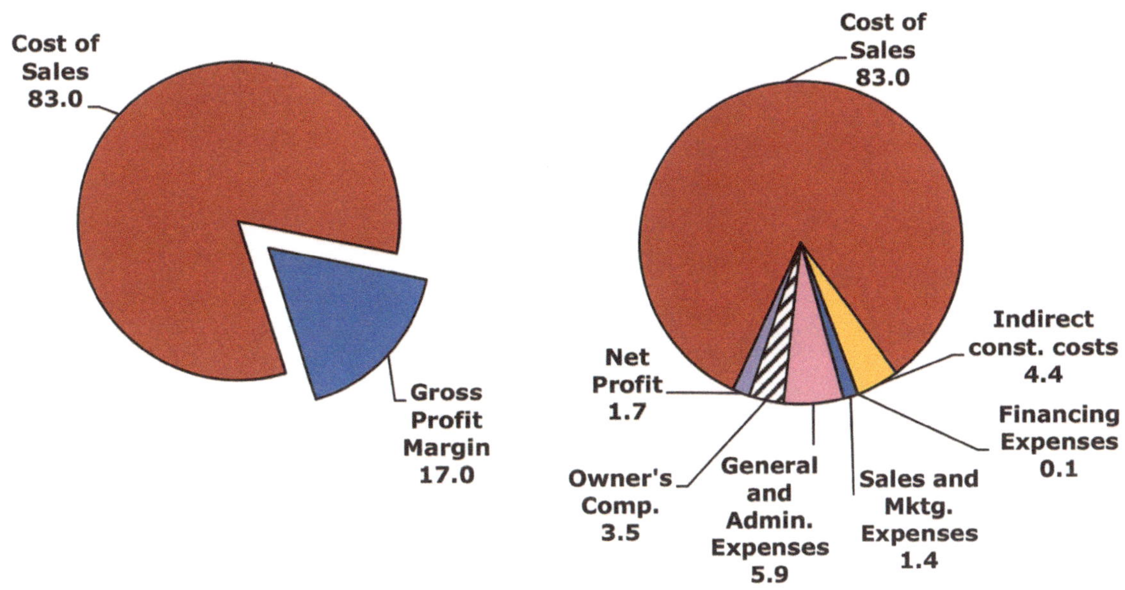

General remodelers had an average of $421,000 worth of assets on their 2021 balance sheets, about 29% more than in 2018 ($326,000) but 6% less than in 2015 ($446,000). During this period, general remodelers have shifted the composition of their assets away from accounts and notes receivables into cash. In 2015, cash accounted for 21.9% of all their assets, a much lower share than the 32.9% in 2018, and 44.4% in 2021. Receivables, on the other hand, declined from representing 35.9% of assets in 2015 to only 20.7% in 2021 (table 2.3).

On the liability front, general remodelers increased the average amount of money they owed by 56% between 2018 ($167,000) and 2021 ($261,000). A noteworthy development is the ongoing move away from financing operations through long-term liabilities and use of current debt instead. In 2015, for example, 35.5% of general remodelers' assets were financed through current liabilities; by 2021, that share was 50.2%. Meanwhile, owner's equity averaged $160,000 in 2021, essentially unchanged from 2018 ($158,000). As a result, equity only financed 37.9% of assets in 2021, compared to 48.6% in 2018.

Table 2.3. Balance Sheet

	2015		2018		2021	
	Average ($1,000s)	**Share of total (%)**	**Average ($1,000s)**	**Share of total (%)**	**Average ($1,000s)**	**Share of total (%)**
Assets						
Current assets						
Cash	98	21.9	107	32.9	187	44.44
Accounts/notes receivable	160	35.9	83	25.4	87	20.7
Other current assets	126	28.3	65	19.8	96	22.8
Total current assets	384	86.1	255	78.2	370	87.9
Other assets	62	13.9	71	21.8	51	12.1
Total assets	**446**	**100.0**	**326**	**100.0**	**421**	**100.0**
Liabilities						
Current liabilities	158	35.5	127	38.8	211	50.2
Long term liabilities	114	25.5	41	12.5	50	11.9
Total liabilities	**272**	**61.0**	**167**	**51.4**	**261**	**62.1**
Owner's equity	**174**	**39.0**	**158**	**48.6**	**160**	**37.9**
Total liabilities & owner's equity	**446**	**100.0**	**326**	**100.0**	**421**	**100.0**

The combination of a much lower average net profit margin along with higher debt levels caused significant deterioration in general remodelers' financial ratios. The average current ratio dropped to 1.75, down from 2.01 in 2018 and 2.43 in 2015. Meanwhile, their level of risk increased, as the average debt-to-equity ratio rose to 1.65, up from 1.06 in 2018 and 1.56 in 2015. Additionally, their ability to turn assets and equity into profits continued to worsen, as the ROA fell to 7.0% (from 24.8% in 2018 and 25.5% in 2015) and the ROE went to 18.6% (from 51.0% in 2018 and 65.3% in 2015) (table 2.4).

Table 2.4. Financial Ratios

	2015	**2018**	**2021**
Current ratio	2.43	2.01	1.75
Debt-to-equity ratio	1.56	1.06	1.64
Return on assets (%)	25.5	24.8	7.0
Return on equity (%)	65.3	51.0	18.6

Note: No analysis for the top and bottom 25 percent of general remodelers was produced due to insufficient sample size.

3

Design-Build Remodelers

Fifty-one percent of the residential remodelers in the study identified as being primarily design-build remodelers. Such companies work with the homeowner to provide design and construction services under a single contract. This chapter analyzes the financial performance of this subset of respondents.

Balance Sheet and Income Statement

The average design-build remodeler took in $1.8 million in revenue in 2021, 36% less than in 2018 ($2.9 million), but the same as in 2015 ($1.8 million). Their income statements also show the overall cost of sales (relative to income) changed very little between 2018 (68.4%) and 2021 (69.0%). During this period, most cost items accounted for a lower or similar share of revenue, with one exception: trade contractors rose from costing 26.8% of revenue in 2018 to 34.4% in 2021. Given the overall stability in the share of revenue spent on cost of sales, design-build remodelers saw their gross profit margin remain largely unchanged from 2018 (31.6%) to 2021 (31.0%) (table 3.1).

Table 3.1 Gross Profit

	2015		2018		2021	
	Average ($1,000s)	Share of revenue (%)	Average ($1,000s)	Share of revenue (%)	Average ($1,000s)	Share of revenue (%)
Revenue	1,831	100.0	2,880	100.0	1,832	100.0
Cost of sales						
Labor for remodeling/repairs	259	14.2	449	15.6	213	11.6
Materials for remodeling/repairs	364	19.9	532	18.5	304	16.6
Trade contractors for remodeling/repairs	544	29.7	773	26.8	630	34.4
Direct costs: commercial remodeling/repairs	23	1.3	4	0.2	6	0.3
Direct costs: single-family and Land	92	5.0	111	3.8	63	3.4
Other costs	25	1.3	102	3.5	49	2.7
Total cost of sales	1,308	71.4	1,971	68.4	1,264	69.0
Gross profit	**524**	**28.6**	**910**	**31.6**	**567**	**31.0**

In addition to keeping their total cost of sales in check, design-build remodelers were also able to bring down their operating expenses in 2021. Altogether, those expenses took up 24.2% of revenue, a whole two points lower than in 2018 (26.2%) (table 3.2). Declines in general and administrative expenses (13.0% to 8.4%) as well as sales and marketing expenses (4.2% to 1.9%) are responsible for the overall drop in operating expenses.

Table 3.2 Net Profit

	2015		2018		2021	
	Average ($1,000s)	Share of revenue (%)	Average ($1,000s)	Share of revenue (%)	Average ($1,000s)	Share of revenue (%)
Gross profit	524	28.6	910	31.6	567	31.0
Operating expenses						
Indirect construction costs	69	3.8	96	3.3	118	6.4
Financing expenses	5	0.3	4	0.1	24	1.3
Sales & mktg. expenses	55	3.0	120	4.2	35	1.9
General & administrative expenses	195	10.7	375	13.0	153	8.4
Owner's compensation	104	5.7	159	5.5	114	6.2
Total operating expenses	428	23.4	754	26.2	444	24.2
Net profit (loss)	**95**	**5.2**	**155**	**5.4**	**123**	**6.7**

In the end, this group of remodelers reported an average of $123,000 in net profits for 2021, a 6.7% net profit margin. The latter is higher than their net margin in 2018 (5.4%), and in fact, is the highest net margin reported by this subset of remodelers since first calculated in 2011. Figure 3.1 summarizes the income statement for design-build remodelers for 2021.

Meanwhile, the balance sheet of the average design-build remodeler in 2021 showed $526,000 in total assets, 15% higher than in 2018 ($459,000) and 28% more than in 2015 ($411,000). The breakdown of their assets did not change substantially from 2018 to 2021, when 44.1% was held in cash, 16.9% in accounts and notes receivables, 10.7% in other current assets, and 28.4% in other fixed assets (table 3.3).

Figure 3.1. **Cost of Sales, Expenses, and Net Profit (% of Revenue)**

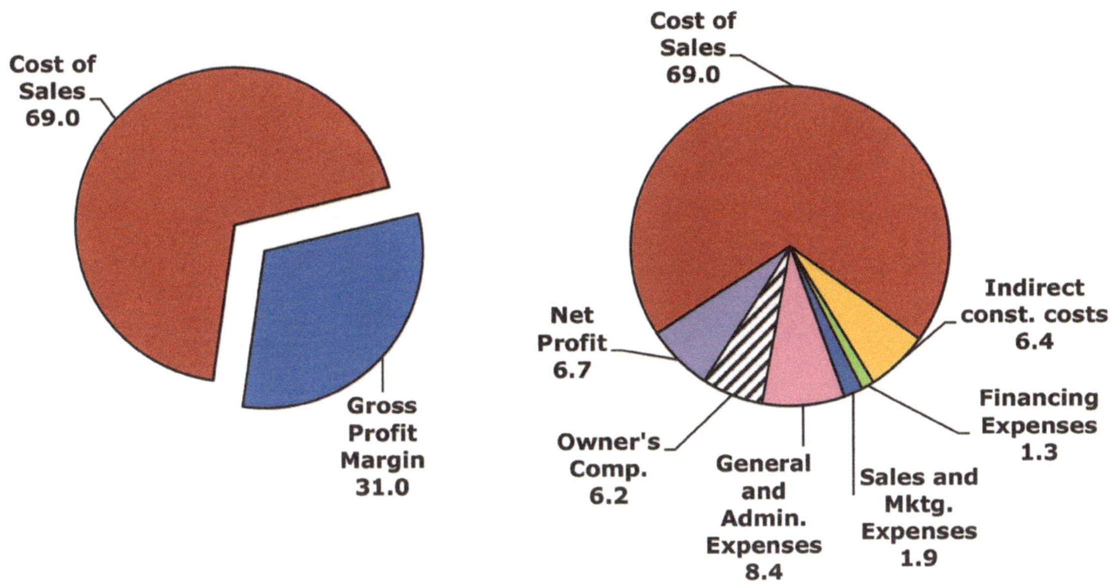

Table 3.3. Balance Sheet

	2015		2018		2021	
	Average ($1,000s)	Share of total (%)	Average ($1,000s)	Share of total (%)	Average ($1,000s)	Share of total (%)
Assets						
Current assets						
Cash	95	23.2	215	46.8	232	44.1
Accounts/notes receivable	113	27.5	74	16.1	89	16.9
Other current assets	68	16.6	49	10.7	56	10.7
Total current assets	276	67.2	338	73.6	377	71.6
Other assets	135	32.8	121	26.4	149	28.4
Total assets	**411**	**100.0**	**459**	**100.0**	**526**	**100.0**
Liabilities						
Current liabilities	241	58.7	206	44.9	133	25.3
Long term liabilities	60	14.7	44	9.5	98	18.6
Total liabilities	**302**	**73.4**	**250**	**54.5**	**231**	**43.9**
Owner's equity	**109**	**26.6**	**209**	**45.5**	**295**	**56.1**
Total liabilities & owner's equity	**411**	**100.0**	**459**	**100.0**	**526**	**100.0**

Further analysis shows that design-build remodelers are not relying on debt to strengthen their balance sheets. In 2021, this group's liabilities averaged $231,000, a lower share of assets (43.9%) than in 2018 (54.5%) or 2015 (73.4%). Instead, their balance sheets are expanding because design-build remodelers are increasing the amount of capital invested in their companies. In 2021, they reported an average of $295,000 in equity, about 40% more than in 2018 ($209,000), and 170% more than in 2015 ($109,000). The latest increase led invested capital to be not only the source of financing for more than half (56.1%) of this group's assets in 2021 (up from 2018's 45.5%), but also the largest share of equity on design-build remodelers' balance sheets since the inception of this series in 2011.

Design-build remodelers' growing balance sheet has strengthened some financial ratios. The combination of higher current assets and lower current liabilities, for example, improved their current ratio from 1.14 in 2015, to 1.64 in 2018, and 2.83 in 2021. That means they had almost three times more liquid assets on hand than short-term debt to pay it with (table 3.4).

Table 3.4. Financial Ratios

	2015	2018	2021
Current ratio	1.14	1.64	2.83
Debt-to-equity ratio	2.76	1.20	0.78
Return on assets (%)	23.2	33.9	23.4
Return on equity (%)	87.3	74.3	41.8

Note: *No analysis for the top and bottom 25 percent of design-build remodelers was produced due to insufficient sample size.*

Fewer debts and more invested capital have also led their debt-to-equity ratio to fall (i.e., improve) from 2.76 in 2015 to 0.78 in 2021. On the other hand, and although still very healthy, their ROA (33.9% in 2018 vs. 23.4% in 2021) and ROE (74.3% vs. 41.8%) have declined because of assets/equity growing at a faster clip than net profit margins.

4

Comparison Across Regions, Business Models, & Historical Trends

This chapter offers readers a chance to look at the results in a different light by doing side-by-side comparisons of profitability and asset benchmarks across regions of the country (only the Midwest and West regions provided enough data to produce results separately) and across the two business models addressed in previous chapters (general remodelers and design-build remodelers). Historical data for all the previous *Remodelers' Cost of Doing Business Studies* produced by the NAHB are also included to identify relevant trends over time.

Revenue & Profits by Region

As mentioned in earlier chapters, residential remodelers responding to this study averaged $1.9 million in total revenue in 2021. As expected, residential remodeling/repair work was the source for most of that revenue (87%), followed by new-single family home building (11%), and to a much lesser extent, commercial remodeling/repairs (2%) and other sources (0.5%).

Remodelers in the West reported an average of $3.1 million in 2021 revenue, nearly twice as much as the $1.6 million reported by remodelers in the Midwest (Appendix A, page A-6). Comparing the sources of revenue between these two regions shows that single-family home building was a significant source of income for remodelers in the West (16%), but not for those in the Midwest (1%). Conversely, nearly all revenue for Midwest remodelers (96%) came from residential remodeling work, compared to a lower 82% for those in the West (fig. 4.1).

Figure 4.1. Regional Differences in Sources of Revenue (% of Revenue)

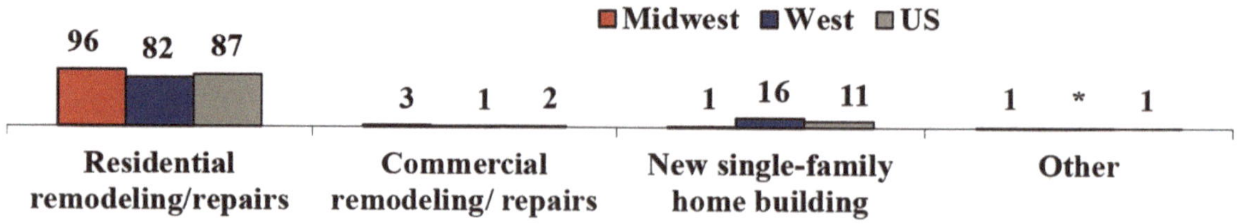

*Less than 0.5%.

In the Midwest, the cost of labor, material, and trade contractors for residential remodeling/repairs took up 62.1% of revenue, not much different from the 63.9% these costs represented for remodelers in the West. The only cost item that varied significantly between these two regions was the share of revenue spent on new single-family home building: only 0.8% for Midwestern remodelers vs. 13.9% for their Western counterparts. After subtracting all costs of sales, remodelers in the Midwest posted a strong average profit margin of 32.9%, compared to a much lower 17.7% among those in the West (fig 4.2).

Figure 4.2. Regional Differences in Costs and Gross Profit Margins (% of Revenue)

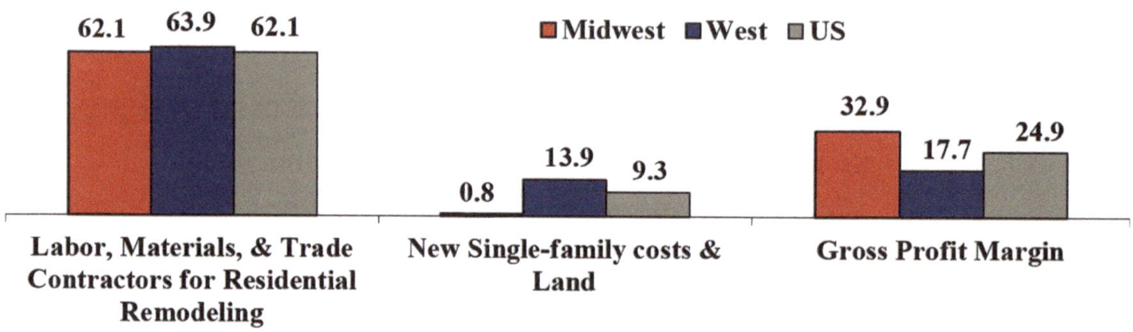

Operating expenses absorbed a much larger share of revenue for remodelers in the Midwest (26.3%) than they did in the West (15.8%). Two expense lines in particular are responsible for the discrepancy: general & administrative expenses, which took up 9.6% of revenue in the Midwest, but only 6.0% in the West; and owner's compensation, which accounted for 6.9% of revenue in the Midwest and 4.0% in the West. In the end, remodelers in the Midwest had a much more profitable year than those in the West, posting average net profit margins of 6.6% vs. 1.9%, respectively. The broader measure of performance that adds together owner's compensation plus the company's net profit margin was 13.5% in the Midwest, more than twice that in the West (5.9%) (fig. 4.3).

Figure 4.3. **Regional Differences in Expenses, Owner's Compensation and Net Profit Margins (% of Revenue)**

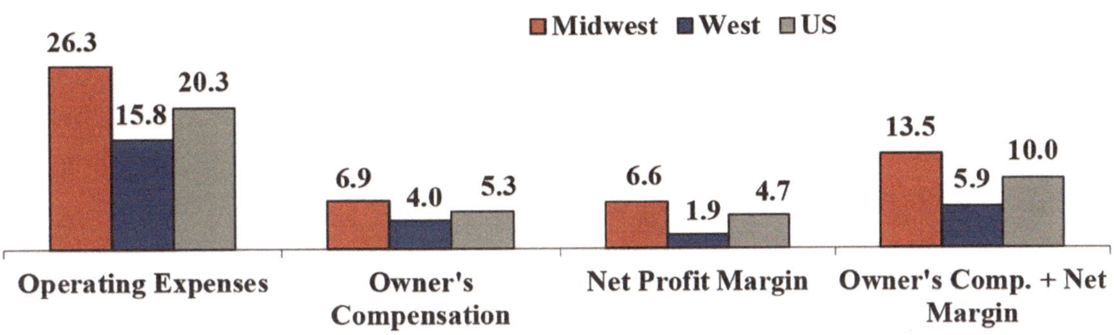

Comparing assets across regions shows that remodelers in the West had the biggest balance sheets, with an average of $737,000 in total assets. In comparison, remodelers in the Midwest reported an average of $474,000 in assets (36% less than those in the West) while those in the South reported $390,000 (47% less) (fig. 4.4). In terms of the composition of those assets, more than 40% of remodelers in every region had over 40% of assets in cash. Meanwhile, the share held as accounts/notes receivables was much larger in the West (28%) and South (24%) than in the Midwest (15%) (Appendix A, page A-4).

Figure 4.4. **Balance Sheet Across Regions ($ Averages in Thousands)**

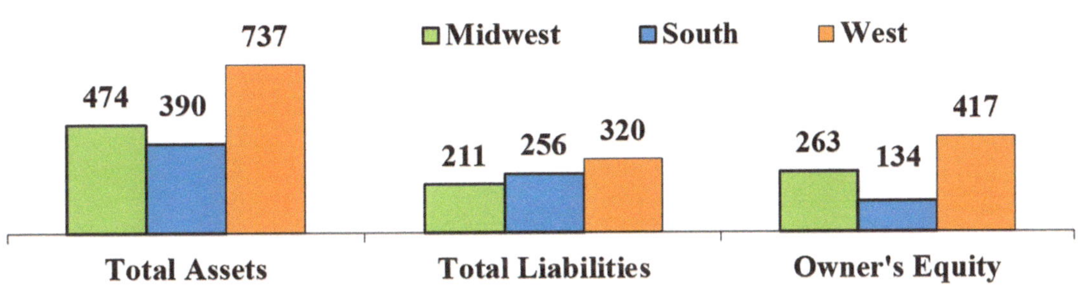

By the end of fiscal year 2021, remodelers in the West owed an average of $320,000 in both current and long-term liabilities, about 25% more than those in the Midwest ($256,000) and 52% more than those in the South ($211,000). Simple arithmetic (assets = liabilities + equity) reveals remodelers in the West had significantly more capital invested in their companies ($417,000) than those in the Midwest ($263,000) or the South ($134,000). These results indicate that equity financed over 55% of all the assets remodelers in both the West and Midwest held in fiscal year 2021. In comparison, invested capital only backed up 34% of the assets held by remodelers in the South.

A comparison of key financial ratios between the three regions (when possible) shows that remodelers in the West outperformed others in debt-related metrics but not in rates of return. For example, the current ratio was highest, at 2.63, in the West, compared to 2.08 in the Midwest and 2.26 in the South. The lowest debt-to-equity ratio (which again measures risk in a company) was also reported by remodelers in the West at 0.77, followed closely by those in the Midwest (0.80) and not so closely by those in the South (1.91).

When it comes to squeezing profits out of assets and equity, however, remodelers in the Midwest were more successful than those in the West. ROA in the Midwest was 22.4%, compared to 7.9% in the West; while their ROE was 40.3% vs. 14.0%, respectively (table 4.1).

Table 4.1. Financial Ratios

	Midwest	South	West
Current ratio	2.08	2.26	2.63
Debt-to-equity ratio	0.80	1.91	0.77
Return on assets (%)	22.4	—	7.9
Return on equity (%)	40.3	—	14.0

Revenue & Profits by Business Model

The *Remodelers' Cost of Doing Business Study* produced results separately for general remodelers and design-build remodelers for the first time in 2011. That year, design-build remodelers posted a higher net profit margin (3.9%) than general remodelers (1.8%). In 2015, the roles were reversed, as general remodelers were somewhat more profitable (6.3%) than design-build remodelers (5.2%). In 2018, the top spot flipped again, with design-build remodelers averaging a 5.4% net margin, slightly ahead of the 4.9% posted by general remodelers. The latest results show further widening in that gap, as design-build remodelers averaged a 6.7% net profit margin in 2021, compared to only 1.7% among general remodelers.

Design-build remodelers averaged $1,832,000 in total revenue in 2021, about 3% more than general remodelers – $1,778,000. Comparing sources of revenue reveals that design-build remodelers generated 94% of it from residential remodeling/repairs and only 4% from new single-family home building. The latter activity, on the other hand, was much more important for general remodelers, bringing in 30% of revenue, while residential remodeling accounted for 67% (Appendix A, page A-6).

The cost of labor, materials, and trade contractors for residential remodeling work took up a larger share of revenue from design-build remodelers (62.6%) than from general remodelers (55.4%), led by higher costs for labor and materials. But the most significant difference in their cost structure was the share of revenue spent on single-family home building: 25.4% among general remodelers vs. only 3.4% among design-build remodelers. Given such variations in costs, the differences in gross margins were staggering: 17.0% among general remodelers compared to 31.0 % among their design-build counterparts (fig. 4.5).

Figure 4.5. Differences in Costs and Gross Profit Margins by Business Model (% of Revenue)

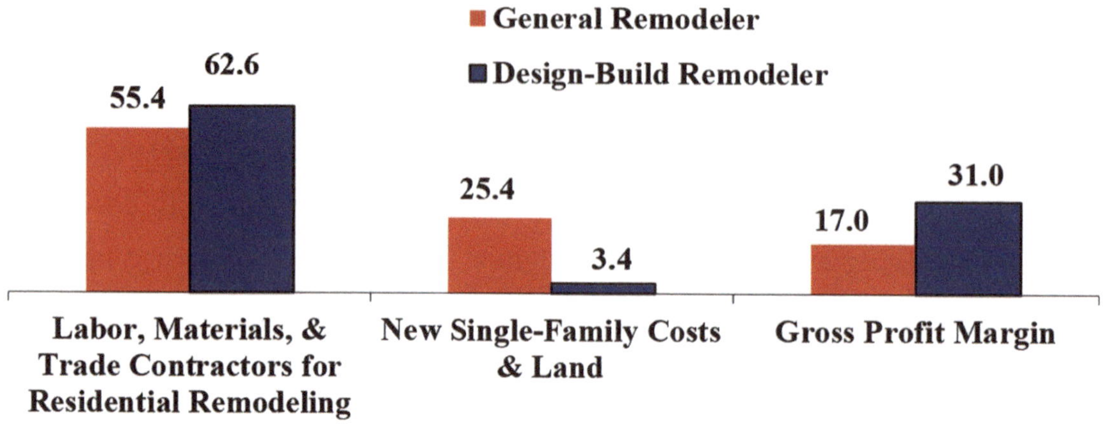

Exactly the opposite of costs of sales, design-build remodelers spent more on operating expenses (24.2% of revenue) than did general remodelers (15.3%). Two expense lines are most responsible for the discrepancy: design-build remodelers spent more paying themselves (6.2%) and on general & administrative expenses (8.4%) than did general remodelers (3.5% and 5.9%, respectively). In the end, design-build remodelers had a much more profitable year than general remodelers, posting average net profit margins of 6.7% vs. 1.7%, respectively. Looking at the broader measure of profitability that combines owner's compensation and net margin together confirms the large gap in performance: 12.9% among design-build remodelers vs. 5.2% among general remodelers (fig. 4.6). **[Perspective from a Remodeler:** Hypothetically removing the revenue and expenses generated by new single-family home building from general remodelers' income statements would bring the two business models much closer together in performance. With 30% of their revenue coming from this activity, likely done at very low margins due to elevated construction costs, it is not surprising to see general remodelers' overall gross profit margin fall to 17%.]

Figure 4.6. **Differences in Expenses, Owner's Compensation and Net Profit Margins by Business Model (% of Revenue)**

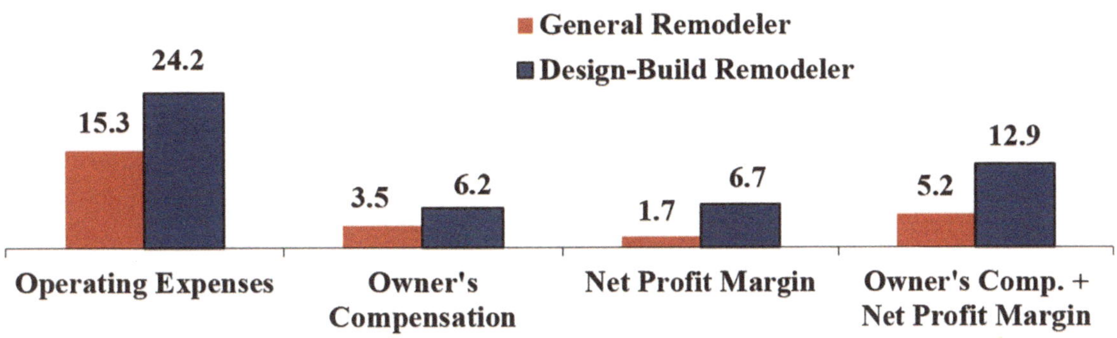

Design-build remodelers had $526,000 worth of assets on their 2021 balance sheets, about 25% more than general remodelers — $421,000 (fig. 4.7). Although the two groups held similar shares of their assets in cash (44%), overall, they had different shares held as current assets (cash + accounts receivables + other current assets): 88% for general remodelers vs. 72% for design-build remodelers (Appendix A, page A-4). **[Perspective from a Remodeler:** General remodelers had far less fixed assets than their design-build counterparts, which could indicate either some of their fixed assets were fully depreciated and would soon need replacing or they generally avoid buying buildings, trucks, and tools.]

Figure 4.7. **Balance Sheet by Business Model ($ Averages in Thousands)**

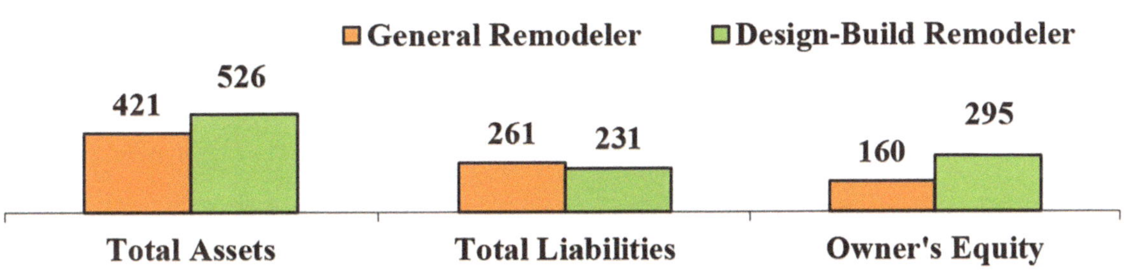

General remodelers owed an average of $261,000 in current and long-term liabilities and had $160,000 worth of equity invested in their companies. These numbers imply that 62% of their assets were financed through debt and the other 38% through equity. Meanwhile, design-build remodelers owed an average of $231,000 and had $295,000 of their own money invested, which means that 44% of the assets on their books were owed to someone else and 56% were owned free and clear.

A comparison of key financial ratios between the two business models shows that design-build remodelers outperformed their general counterparts in all metrics. For example, the former had a higher current ratio (2.83) than the latter (1.75). The level of risk, measured by the debt-to-equity ratio, was lower for design-build remodelers, at 0.78, than for general remodelers, at 1.64. And when it comes to turning assets and equity into profits, design-build remodelers also outmaneuvered their general counterparts: their respective ROAs were 23.4% vs. 7.0% and ROEs were 41.8% vs. 18.6% (table. 4.2).

Table 4.2. Financial Ratios

	General Remodelers	Design-build Remodelers
Current ratio	1.75	2.83
Debt-to-equity ratio	1.64	0.78
Return on assets (%)	7.0	23.4
Return on equity (%)	18.6	41.8

Historical Trends

The *Remodelers' Cost of Doing Business Study* has been conducted periodically since 1990. That year, the average net profit margin for this segment of the industry was reported at 11.4% — the highest in the series. In both 1993 and 1996, remodelers reported an average net profit margin of 6.8%, but in the next two iterations of the study, in 2003 and 2011, the net margin declined to 4.2% and 3.0%, respectively. The downward trend was reversed in

2015, however, when it rose to 5.3%, and then remained essentially unchanged in 2018, at 5.2%. By 2021, a pandemic-driven boom in housing demand led some remodelers to enter the new single-family home building business, which led to disproportionate increases in their cost structure. That, coupled with a steep increase in the cost of trade contractors for residential remodeling work led remodelers' average net profit margin to drop to 4.7% in 2021 (table 4.3).

Throughout much of the history of this study (from 1993 to 2015), remodelers have spent between 71% and 73% of their revenue on the various items under costs of sales (e.g., labor, material, and trade contractors). In 2018, costs fell below that range, taking up 69.9% of revenue, but then jumped to an all-time high of 75.1% in 2021. As a result, remodelers' gross margin reached its lowest point in the history of this series that year, at 24.9%.

Operating expenses were highest in 1990, when they accounted for 26.7% of all revenue. By 1993, remodelers had greatly tightened the belt around these expenses, driving them down to only 21.9% of revenue. From that year until 2018, they accounted for somewhere between 22% and 25% of revenue. But in 2021, remodelers managed to shrink them even further, down to only 20.3% – the lowest share in the history of the series.

Table 4.3. Historical Comparison of Profitability (% of Revenue)

	Total Revenue	Cost of Sales	Gross Profit Margin	Operating Expenses	Net Profit Margin
2021	100	75.1	24.9	20.3	4.7
2018	100	69.9	30.1	24.8	5.2
2015	100	71.1	28.9	23.6	5.3
2011	100	73.2	26.8	23.7	3.0
2003	100	71.6	28.4	24.3	4.2
1996	100	71.0	29.0	22.2	6.8
1993	100	71.3	28.7	21.9	6.8
1990	100	61.9	38.1	26.7	11.4

Figure 5.3. **Years in Business (% of Respondents)**

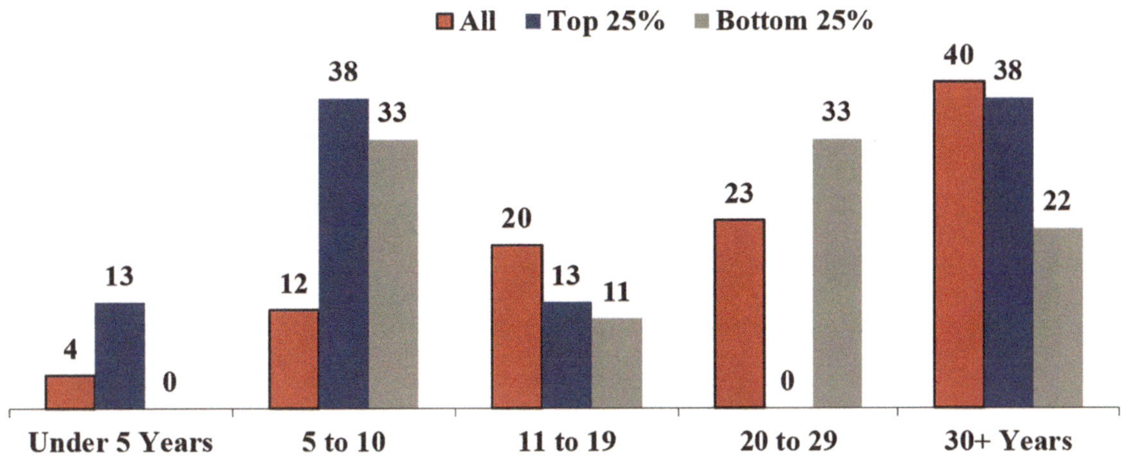

On average, remodelers in the most successful quartile have been in business for 24 years, whereas those in the bottom quartile have been in business for 21 years (Appendix B, page B-4). In fact, a more telling statistic of the value of longevity is that 38% of the top 25% group have been in business for 30+ years, compared to only 22% among those in the bottom 25% group.

Looking at profits by tenure shows that remodelers with more experience (over 20 years in the business) had essentially the same gross profit margin (24.9%) as those with relatively 'less' experience (24.7%). The two groups differed on how much revenue each spent on operating expenses, however: the group with 20 or fewer years in the business spent 18.3%, whereas those with more years of experience spent 22.6%. Most of the difference came in the form of higher indirect costs (3.7% vs. 5.9%, respectively). As a result, the companies in business for 20 years or less outperformed their older counterparts in net profit margins: 6.4% vs. 2.3% (Appendix A, page A-6).

Number of Jobs Completed

Residential remodelers were also asked about the number of jobs completed during 2021 in each of four different price ranges. On average, remodelers completed 19 jobs priced under $5,000, 16 jobs priced between $5,000 and $24,999, 17 jobs priced between $25,000 and $99,999, and 10 jobs worth $100,000 or more. Altogether, remodelers responding to the survey completed an average of 61 jobs in 2021 (fig. 5.4).

Figure 5.4. Average Number of Jobs Completed in 2021 by Price Range

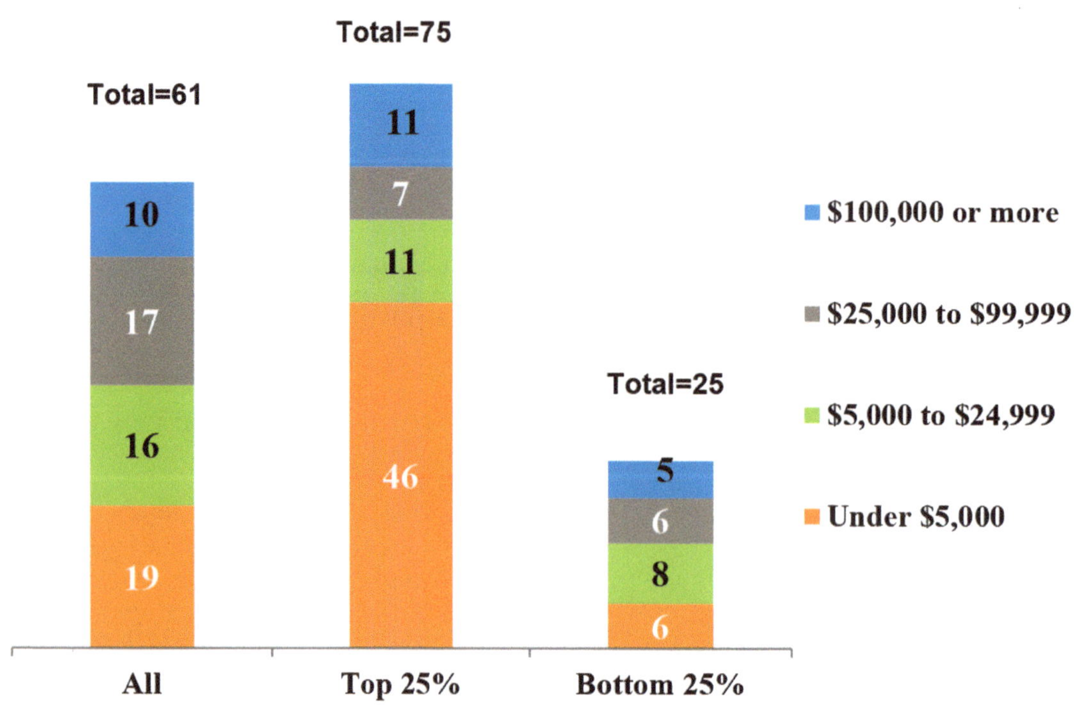

Remodelers with net profit margins in the top quartile completed an average of 75 jobs in 2021, the majority of which (46) were under $5,000 and 11 were $100,000 or more. Remodelers in the lowest quartile, in contrast, completed an average of only 25 jobs, one-third the number of the most successful group. Those jobs were about equally distributed across the different price ranges, with 6 being priced under $5,000 and 5 worth $100,000 or more. **[Perspective from a Remodeler:** It appears that remodelers doing multiple, similar size jobs, likely of similar scopes, have a much greater chance of turning repeatable profits. One possible explanation for this is that they get more frequent looks at their costs, which allows them to tweak their pricing more contemporaneously.]

An analysis of the number of jobs completed by business model shows that, on average, design-build remodelers finished 53 jobs in 2021, 20 fewer than the 73 finished by general remodelers. Regionally, respondents in the Northeast completed an average of 42 jobs in 2021, compared to 59 in the South, 62 in the Midwest, and 75 in the West. Remodelers with more than 20 years in the business finished 68 jobs, compared to 45 among those with less tenure (Appendix A, page A-3).

Meanwhile, profitability and the number of jobs completed showed a direct relationship in 2021. Remodelers who finished 30 or fewer jobs posted a net profit margin of 3.4%, compared to 6.0% among those who finished more than 30 jobs (Appendix A, page A-6). Results also show a significant difference in the balance sheets of these two groups: remodelers with more than 30 jobs completed reported an average of $715,000 in total assets, nearly double the $359,000 average among those with up to 30 job completions (Appendix A, page A-4).

Number of Full-Time Employees

Finally, the survey inquired about the average number of full-time employees the company had on payroll in 2021. The question instructed the respondent to include himself or herself in the count, and if the firm had any part-time employees, to use full-time equivalents (that is, number of paid hours divided by 2,080 hours).

Residential remodelers who participated in the study had an average of 7.8 employees on payroll in 2021. Of them, 4.3 worked in the field, 2.4 worked in the office, and 1.1 in sales. In all, 24% reported having up to 2.0 full-time positions (or equivalents), another 18% had 2.1 to 4 positions, 17% had 4.1 to 6.0, 16% had 6.1 to 10, another 16% had 10.1 to 15, and 10% had more than 15 employees on payroll (fig. 5.5)

Figure 5.5. **Number of Full-Time Employees on Payroll in 2021 (% of Respondents)**

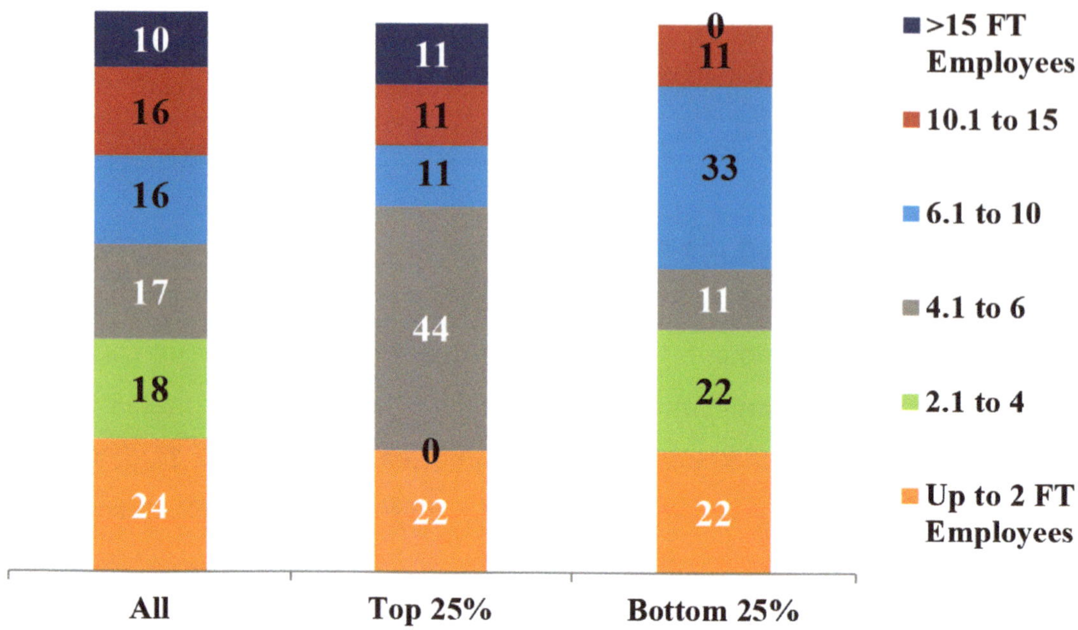

Looking at the average number of employees across the top and bottom 25% quartiles shows that both groups had about the same number of full-time positions in 2021: 6.4 among the most successful group vs. 6.2 among the least successful one (Appendix B, page B-1). Across regions, remodelers in the West had the most full-time positions, at 9.3, compared to 6.8 in the South, 7.1 in the Northeast, and 8.6 in the Midwest (Appendix A, page A-3).

Design-build remodelers had 9.1 full-time employees in 2021, compared to 6.2 among general remodelers. Those with more than 20 years in business reported an average of 8.9 employees, compared to 5.9 among those with up to 20 years as remodelers. Meanwhile, remodelers whose revenue in 2021 was under $2 million reported an average of 3.9 full-time positions, compared to 12.0 among those earning $2 million or more (Appendix A, page A-3).

Appendix

A

Detailed Tables

Q1a. Indicate your firm's main operation during Fiscal 2021
(Percent of Respondents)

	Total	Region				Type of Residential Remodeler		Total 2021 Revenue		Number of Jobs		Number of Years in Business	
		Northeast	Midwest	South	West	General	Design-build	< $2 million	≥ $2 million	≤ 30	> 30	≤ 20	> 20
Residential remodeling/rehabilitation	85	96	91	80	82	100	100	100	100	100	100	100	100
Commercial remodeling/rehabilitation	1			1	2								
Single-family home building	12	4	7	16	14								
Multifamily home building	0			1									
Other	2		2	2	2								

Q1b. Percent of your business from each operation
(Average Share)

	Total	Region				Type of Residential Remodeler		Total 2021 Revenue		Number of Jobs		Number of Years in Business	
		Northeast	Midwest	South	West	General	Design-build	< $2 million	≥ $2 million	≤ 30	> 30	≤ 20	> 20
Residential remodeling/rehabilitation	78	89	83	74	76	89	90	92	90	90	89	89	91
Commercial remodeling/rehabilitation	5	5	5	5	5	5	3	5	3	3	6	4	4
Single-family home building	14	6	11	18	14	4	6	2	5	6	4	6	3
Multifamily home building	1	0	1	1	1	1	0	0	1	0	0	0	0
Other	2	0	1	2	5	2	0	0	1	1	1	1	1

Q2. If a residential remodeler, please indicate best description of the firm
(Percent of Respondents)

	Total	Region				Type of Residential Remodeler		Total 2021 Revenue		Number of Jobs		Number of Years in Business	
		Northeast	Midwest	South	West	General	Design-build	< $2 million	≥ $2 million	≤ 30	> 30	≤ 20	> 20
General remodeler	42	55	27	51	36	100		39	15	39	44	43	43
Design-build remodeler	51	45	65	42	56		100	57	77	51	52	45	52
Specialty contractor	4		6	5	3			4		6	2	9	2
Repair & maintenance contractor	3		2	2	6				8	3	2	3	3

Q3. Years firm has been in residential remodeling business.
(Percent of Respondents)

	Total	Region				Type of Residential Remodeler		Total 2021 Revenue		Number of Jobs		Number of Years in Business	
		Northeast	Midwest	South	West	General	Design-build	< $2 million	≥ $2 million	≤ 30	> 30	≤ 20	> 20
Under 5 years	4	10		4	6	7	2	10		7	1	10	
5-10 years	12	10	14	16	3	11	11	33	10	14	10	29	
11-19 years	20	19	18	22	22	17	22	19	10	24	18	49	
20-29 years	23	5	20	28	28	24	22	10	40	19	25	12	31
30+ Years	40	57	48	30	41	42	42	29	40	36	46		69
Average (Years)	26	27	29	23	27	26	26	20	30	24	28	12	36

Q4. How many jobs in each of these price ranges did your firm complete in 2021?
(Average Number of Jobs)

	Total	Region				Type of Residential Remodeler		Total 2021 Revenue		Number of Jobs		Number of Years in Business	
		Northeast	Midwest	South	West	General	Design-build	< $2 million	≥ $2 million	≤ 30	> 30	≤ 20	> 20
Under $5,000	19	9	15	19	29	31	10	21	6	3	37	14	21
$5,000 to $24,999	16	16	20	14	13	12	19	8	10	4	29	11	18
$25,000 to $99,999	17	12	17	19	13	23	10	7	7	3	32	14	19
$100,000 or more	10	5	10	7	20	7	14	4	12	4	17	7	10
Average # of jobs	61	42	62	59	75	73	53	40	35	14	115	45	68
Median # of jobs	27	27	38	25	29	29	29	16	31	14	71	26	37

Q5. On avearge, how many full-time employees (including yourself) did your firm have on payroll in 2021?
(Percent of Respondents)

	Total	Region				Type of Residential Remodeler		Total 2021 Revenue		Number of Jobs		Number of Years in Business	
		Northeast	Midwest	South	West	General	Design-build	< $2 million	≥ $2 million	≤ 30	> 30	≤ 20	> 20
Average # of Employees													
Office	2.4	2.1	2.9	1.8	3.2	1.5	3.1	1.2	4.4	1.6	3.2	1.7	2.7
Field	4.3	4.0	4.1	4.3	4.6	3.9	4.6	2.4	6.5	2.4	6.3	3.4	4.8
Sales	1.1	0.9	1.6	0.7	1.5	0.8	1.4	0.3	1.1	0.5	1.7	0.7	1.3
Total Average	7.8	7.1	8.6	6.8	9.3	6.2	9.1	3.9	12.0	4.5	11.3	5.9	8.9
Percent of Respondents													
Zero employees	3	5	7		3	4	2	4		5		6	1
0.1 to 2	21	18	13	27	21	24	15	35		34	6	31	14
2.1 to 4	18	32	17	16	12	21	13	26		25	10	13	20
4.1 to 6	17	14	13	19	21	16	20	22		13	22	19	16
6.1 to 10	16	9	17	17	15	12	20	4	46	15	17	13	17
10.1 to 15	16	9	20	16	15	19	16	9	38	5	28	10	19
More than 15 employees	10	14	13	5	15	4	13		15	2	18	6	14
Median	5	4	7	5	6	5	6	4	12	3	10	4	6

Q6. Balance Sheet
(Average in $1,000s)

	Total	Region				Type of Residential Remodeler		Total 2021 Revenue		Number of Jobs		Number of Years in Business	
		Northeast	Midwest	South	West	General	Design-build	< $2 million	≥ $2 million	≤ 30	> 30	≤ 20	> 20
Assets													
Cash	210		205	171	299	187	232	76	363	143	312	100	268
Accounts and notes receivables	113		71	92	209	87	89	37	259	85	168	17	161
Other current assets	69		84	26	101	96	56	72	115	66	68	25	116
Other assets	105		113	101	127	51	149	29	234	65	167	32	137
Total assets	**497**		**474**	**390**	**737**	**421**	**526**	**214**	**972**	**359**	**715**	**173**	**681**
Liabilities													
Current liabilities	165		173	128	231	211	133	62	393	154	148	41	244
Long-term liabilities	77		38	128	88	50	98	46	58	77	78	25	118
Owner's Equity	255		263	134	417	160	295	106	521	127	488	108	318
Total Liabilities + Equity	**497**		**474**	**390**	**737**	**421**	**526**	**214**	**972**	**359**	**715**	**173**	**681**

Q7. Income Statement
(Average in $1,000s)

	Total	Region				Type of Residential Remodeler		Total 2021 Revenue		Number of Jobs		Number of Years in Business	
		Northeast	Midwest	South	West	General	Design-build	< $2 million	≥ $2 million	≤ 30	> 30	≤ 20	> 20
Revenue													
Residential remodeling/repairs	1,667		1,543		2,561	1,198	1,717	594	3,564	1,289	2,261	1,006	1,968
Commercial remodeling/repairs	32		41		38	45	28	46	7	29	37	45	30
New single-family home building	211		14.48		507	531	76	88	429	266	125	33	96
All other revenue	9		11		4	4	11	5	16	13	3	-	13
Total company revenue	1,919		1,610		3,111	1,778	1,832	734	4,015	1,596	2,426	1,085	2,107
Cost of Sales													
Labor for residential remodeling/repairs (inc. fringe benefits)	209		195		367	123	213	104	394	95	388	197	235
Materials for residential remodeling/repairs	300		302		453	221	304	148	569	200	457	231	346
Trade Contractors for residential remodeling/repairs	682		503		1,169	641	630	188	1,557	641	748	324	781
Commercial remodeling and repairs direct construction costs	12		18.907		11	26	6	17	3	12	11	9	19
New single-family direct construction costs and land	178		12.509		433	452	63	69	372	226	103	33	91
All other costs of sales	59		49		127	12	49	12	143	41	89	22	111
Total company cost of sales	1,440		1,080		2,560	1,476	1,264	537	3,038	1,214	1,796	817	1,583
Gross Profit	**479**		**530**		**551**	**302**	**567**	**196**	**978**	**382**	**630**	**268**	**524**
Operating Expenses													
Indirect construction costs	102		91		148	79	118	25	237	94	114	40	125
Financing expenses	16		36		0	1	24	2	40	25	1	3	33
Sales and marketing expenses	30		31		32	24	35	15	58	32	27	17	32
General and administrative expenses	139		154		187	106	153	60	278	100	200	79	165
Owner's compensation	102		111		125	62	114	48	199	78	141	59	120
Total operating expenses	389		424		493	272	444	150	811	328	484	199	475
Net Income Before Taxes	**90**		**106**		**58**	**30**	**123**	**46**	**166**	**54**	**146**	**70**	**49**

Q7. Income Statement
(Share of Revenue)

	Total	Region				Type of Residential Remodeler		Total 2021 Revenue		Number of Jobs		Number of Years in Business	
		Northeast	Midwest	South	West	General	Design-build	< $2 million	≥ $2 million	≤ 30	> 30	≤ 20	> 20
Revenue													
Residential remodeling/repairs	86.9		95.8		82.3	67.4	93.7	81.0	88.8	80.7	93.2	92.8	93.4
Commercial remodeling/repairs	1.7		2.6		1.2	2.5	1.5	6.3	0.2	1.8	1.5	4.2	1.4
New single-family home building	11.0		0.9		16.3	29.9	4.2	12.0	10.7	16.6	5.2	3.1	4.6
All other revenue	0.5		0.7		0.1	0.2	0.6	0.7	0.4	0.8	0.1	0.0	0.6
Total company revenue	100.0		100.0		100.0	100.0	100.0	100.0	100.0	100.0	100.0	100.0	100.0
Cost of Sales													
Labor for residential remodeling/repairs (inc. fringe benefits)	10.9		12.1		11.8	6.9	11.6	14.2	9.8	5.9	16.0	18.2	11.2
Materials for residential remodeling/repairs	15.6		18.7		14.6	12.4	16.6	20.1	14.2	12.5	18.8	21.3	16.4
Trade contractors for residential remodeling/repairs	35.6		31.2		37.6	36.1	34.4	25.6	38.8	40.1	30.8	29.9	37.1
Commercial remodeling and repairs direct construction costs	0.6		1.2		0.4	1.5	0.3	2.3	0.1	0.8	0.4	0.8	0.9
New single-family direct construction costs and land	9.3		0.8		13.9	25.4	3.4	9.4	9.3	14.2	4.2	3.1	4.3
All other costs of sales	3.1		3.0		4.1	0.7	2.7	1.6	3.6	2.5	3.7	2.0	5.3
Total company cost of sales	75.1		67.1		82.3	83.0	69.0	73.2	75.7	76.1	74.0	75.3	75.1
Gross Profit	**24.9**		**32.9**		**17.7**	**17.0**	**31.0**	**26.8**	**24.3**	**23.9**	**26.0**	**24.7**	**24.9**
Operating Expenses													
Indirect construction costs	5.3		5.7		4.8	4.4	6.4	3.5	5.9	5.9	4.7	3.7	5.9
Financing expenses	0.8		2.2		0.0	0.1	1.3	0.3	1.0	1.6	0.0	0.3	1.6
Sales and marketing expenses	1.6		2.0		1.0	1.4	1.9	2.0	1.4	2.0	1.1	1.6	1.5
General and administrative expenses	7.2		9.6		6.0	5.9	8.4	8.2	6.9	6.2	8.3	7.3	7.8
Owner's compensation	5.3		6.9		4.0	3.5	6.2	6.5	5.0	4.9	5.8	5.5	5.7
Total operating expenses	20.3		26.3		15.8	15.3	24.2	20.5	20.2	20.6	20.0	18.3	22.6
Net Income Before Taxes	**4.7**		**6.6**		**1.9**	**1.7**	**6.7**	**6.3**	**4.1**	**3.4**	**6.0**	**6.4**	**2.3**

Appendix

B

Top and Bottom 25% by Net Profit Margin

Income Statement
(Average in $1,000s)

	ALL	Top 25%	Bottom 25%
Revenue			
Residential remodeling/repairs	1,667	1,446	1,478
Commercial remodeling/repairs	32	36	24
New single-family home building	211	0	147
All other revenue	9	8	0
Total company revenue	1,919	1,490	1,649
Costs of sales			
Labor for residential remodeling/repairs	209	187	206
Materials for residential remodeling/repairs	300	173	317
Trade Contractors for residential remodeling/repairs	682	557	628
Commercial remodeling and repairs direct construction costs	12	2	0
New single-family direct construction costs and land	178	0	141
All other costs of sales	59	12	108
Total cost of sales	1,440	932	1,400
Gross profit	**479**	**558**	**249**
Operating expenses			
Indirect construction costs	102	92	151
Financing expenses	16	2	4
Sales and marketing expenses	30	31	40
General and administrative expenses	139	134	114
Owner's compensation	102	81	56
Total operating expenses	389	339	366
Net income before taxes	**90**	**219**	**(117)**

Income Statement
(% share of revenue)

	ALL	Top 25%	Bottom 25%
Revenue			
Residential remodeling/repairs	86.9	97.0	89.6
Commercial remodeling/repairs	1.7	2.4	1.5
New single-family home building	11.0	0.0	8.9
All other revenue	0.5	0.5	0.0
Total company revenue	100.0	100.0	100.0
Costs of sales			
Labor for residential remodeling/repairs	10.9	12.6	12.5
Materials for residential remodeling/repairs	15.6	11.6	19.2
Trade Contractors for residential remodeling/repairs	35.6	37.4	38.1
Commercial remodeling and repairs direct construction costs	0.6	0.1	0.0
New single-family direct construction costs and land	9.3	0.0	8.6
All other costs of sales	3.1	0.8	6.5
Total cost of sales	75.1	62.6	84.9
Gross profit	**24.9**	**37.4**	**15.1**
Operating expenses			
Indirect construction costs	5.3	6.2	9.2
Financing expenses	0.8	0.1	0.2
Sales and marketing expenses	1.6	2.1	2.4
General and administrative expenses	7.2	9.0	6.9
Owner's compensation	5.3	5.4	3.4
Total operating expenses	20.3	22.8	22.2
Net income before taxes	**4.7**	**14.7**	**(7.1)**

Balance Sheet
(Average in $1,000s)

	ALL	Top 25%	Bottom 25%
Assets			
Cash	210	160	112
Accounts and notes receivables	113	58	101
Other current assets	69	91	159
Other assets	105	134	46
Total assets	497	443	418
Liabilities			
Current liabilities	165	112	190
Long-term liabilities	77	66	71
Owner's equity	255	264	157
Total liabilities + Equity	497	443	418

Please indicate best description of the firm
(Percent of Respondents)

	ALL	Top 25%	Bottom 25%
General remodeler	42	22	44
Design-build remodeler	51	78	56
Specialty contractor	4		
Repair & maintenance contractor	3		

Years firm has been in residential remodeling business
(Percent of Respondents)

	ALL	Top 25%	Bottom 25%
Under 5 years	4	13	
5-10 years	12	38	33
11-19 years	20	13	11
20-29 years	23		33
30+ Years	40	38	22
Average (Years)	26	24	21

How many jobs in each of these price ranges did your firm complete in 2021?
(Average Number of Jobs)

	ALL	Top 25%	Bottom 25%
Under $5,000	19	46	6
$5,000 to $24,999	16	11	8
$25,000 to $99,999	17	7	6
$100,000 or more	10	11	5
Average total # of jobs	61	75	25

On avearge, how many full-time employees (including yourself) did your firm have on payroll in 2021?

	ALL	Top 25%	Bottom 25%
Average # of Employees			
Office	2.4	1.9	2.4
Field	4.3	4.0	3.2
Sales	1.1	0.5	0.6
Total average	**7.8**	**6.4**	**6.2**
Percent of Respondents			
Zero employees	3	11	
0.1 to 2	21	11	22
2.1 to 4	18		22
4.1 to 6	17	44	11
6.1 to 10	16	11	33
10.1 to 15	16	11	11
More than 15 employees	10	11	

Appendix C

Glossary

accrual accounting: A system in which item costs are realized at billing and are included in the financial statements as they are earned or incurred rather than as they are received or paid out in cash.

cash budget: A forecasted summary of a firm's expected cash flow and expected cash and loan balances.

completed contract: A procedure for computing payments on a contract in which payment is rendered and, thus, revenue earned after the entire job is satisfactorily completed.

current liabilities: These are a company's financial obligations to its creditors, suppliers, tax authorities, and others. The amount is due within one year.

current ratio: A ratio that reflects the relationship between current assets and current liabilities. It is a simple way to assess liquidity. This ratio provides an indication of a business ability to meet its current liabilities.

Current ratio = total current assets
÷ total current liabilities

debt-to-equityratio: A ratio that provides the measure of money owed to creditors in relation to the equity value of the company.

Debt-to-equity ratio = total liabilities
÷ total equity

depreciation: A decrease in the value of equipment from use and the passage of time. Depreciation on business equipment is generally deductible for tax purposes over a number of years.

fixed asset: A tangible long-term asset such as land, buildings, or machinery, held for use rather than for processing or resale.

gross margin: Pretax net sales minus cost of sales. Also referred to as gross income.

gross profit margin: Gross profit divided by total revenue.

interest expense: Money that is paid out in interest on loans.

liability: A financial obligation (or the cash outlay) that must be made at a specific time to satisfy the contractual terms of such an obligation.

Limited Liability Corporation (LLC): A corporate structure in which the business is liable for any losses or obligations. The business owners are not personally liable.

long-term liabilities: Debts owned by the business that are not due for one or years from the date of the balance sheet.

net income before taxes: Income after cost of sales, operating expenses, sales and marketing expenses, indirect construction expenses, nondeductible deductions, and allowances have been subtracted from gross total income.

operating expenses: Expenses associated with running the business. Listed as a category on the profit and loss statement.

owner's equity: The amount of money that owners have directly invested in the business from the prior year's profits.

partnership: A contract between two or more persons who agree to pool talent and money to share profits or losses.

percentage of completion: An accounting system for computing partial payments on a large contract wherein identifiable portions of the work may be satisfactorily completed, invoiced and paid before the entire project is completed and paid in full. It is the percentage of cost incurred, multiplied by the total sales price for each job. This method can also be used for a single job to better predict actual gross profit at completion.

net profit margin: Net profit before taxes divided by sales for a given 12-month period, expressed as a percentage.

profit sharing: A compensation arrangement in which employees received additional pay or benefits when the company earned or increased profit.

regular (or C) corporation: This is a standard business corporation in which earnings are taxed at the entity profitability level before dividends. It is referred to as a "C" corporation because it is taxed under subsection C of the IRS code.

return on assets: A measure of profitability that indicates how effective the business is at earning money from the total assets dedicated to the business.

$$\text{ROA ratio} = \text{net profit} \div \text{total assets}$$

return on equity: A measure of profitability that indicates how effective the business is at earning money from the owner's invested capital.

$$\text{ROE ratio} = \text{net profit} \div \text{total equity}$$

S corporations: A domestically owned corporation with no more than 75 owners who have elected to pay taxes under subchapter S of the IRS Code. An S Corporation is taxed like a partnership: it is exempt from the corporate income tax, but its owners pay income taxes on all the firm's income, including owner's wages, bonuses, etc., even if some of the earnings are retained by the firm.

sole proprietorship: A sole proprietorship is an unincorporated business that is owned by one individual. The owner is personally liable for any business debts.

Appendix

D

2023 Remodelers' Cost of Doing Business Survey

Please report on your **fiscal year 2021 operations**. Participants will receive a **FREE electronic copy** of the *Remodelers' Cost of Doing Business Study, 2023 Edition*.

I. FIRM'S PROFILE

1. Indicate your firm's main operation during fiscal year 2021 and the percentage of your business each operation represents.

	Main operation (Check ONE only)	Percent of Your Business
Residential remodeling/rehabilitation	☐	_______ %
Commercial remodeling/rehabilitation	☐	_______ %
Single-family home building	☐	_______ %
Multifamily home building	☐	_______ %
Other	☐	_______ %

2. If your firm's main operation is residential remodeling/rehabilitation, please indicate which of the following best describes your firm:

☐ General remodeler ☐ Specialty contractor
☐ Design-build remodeler ☐ Repair & maintenance contractor

3. How many years has your firm been in the residential remodeling business? ______

4. How many jobs in each of the following price ranges did your firm complete in 2021?

Under $5,000 _________ $25,000 to $99,999 _________
$5,000 to $24,999 _________ $100,000 or more _________

5. On average, how many full-time employees (including yourself) did your firm have on payroll in 2021? (If you had part-time employees, please use full-time equivalents, defined as number of paid hours in each job class divided by 2,080 hours)

Office ____________ Field _____________ Sales _____________

II. FINANCIAL INFORMATION

6. Balance Sheet

Each code number refers to the NAHB Chart of Accounts, Appendix D, found at www.nahb.org/chart

Assets		Liabilities and Owner's Equity	
Cash [1000-1090]	$	**Current liabilities** [2000-2490]	$
Accounts and notes receivables [1200-1290]	$	**Long-term liabilities** [2500-2700]	$
Other current assets [1100-1190; 1300-1690]	$	**Owner's equity** [2900-2960]	$
Other assets [1700-1990]	$		
Total Assets*	$	**Total Liabilities and Owner's Equity***	$

* Total assets must equal the total of liabilities plus owner's equity.

7. Income Statement

*Each code number refers to the NAHB Chart of Accounts, Appendix D, found at www.nahb.org/chart		
Revenue from Operations		
Residential remodeling and repairs [3130,3135,3137]	$	
Commercial remodeling and repairs [3133]	$	
New single-family home building [3100-3125]	$	
All other revenue [3000, 3050, 3140, 3150, 3160-3490]	$	
Total company revenue [sum of the previous four entries]		$
Cost of Sales		
Labor for residential remodeling/repairs (include fringe benefits) [3810-3820]	$	
Material for residential remodeling/repairs [3830]	$	
Trade contractors for residential remodeling/repairs [3840]	$	
Commercial remodeling and repairs direct construction costs [3633]	$	
New single-family direct construction costs [3600,3610,3620,3625] (permits, labor with burden, materials, trade contractors) and land [3550]	$	
All other costs of sales (related to items that fall under "All other revenue")	$	
Total company cost of sales [sum of previous six entries]		$
Gross Profit		
Total Company Revenue minus Total Company Cost of Sales		$
Operating Expenses		
Indirect construction costs [4000-4990] (Superintendents, vehicles, tools, warranty, construction equipment)	$	
Financing expenses [5000-5990] (points and interest on all business loans)	$	
Sales and marketing expenses [6000-6990] (advertising, marketing, salaries and commissions to sales personnel)	$	
General and administrative expenses [8000-8990] (salaries, payroll taxes and benefits, office expenses, vehicles, travel, entertainment, taxes, insurance, professional services, and depreciation) (exclude owner's compensation)	$	
Owner's compensation [8010] (owner's salary, draws, bonuses and benefits. Do not include distributions from retained earnings)	$	
Total operating expenses [sum of previous five entries]		$
Net Income Before Taxes [total company gross profit minus total operating expenses]		$

8. Comments: ______________________________

Please provide email where you would like to receive your free copy of the *Remodelers' Cost of Doing Business, 2023 Edition:* ______________________________